BASIC MATH WITH A CALCULATOR

South-Western

William R. Pasewark
Professor Emeritus
Texas Tech University
Lubbock, Texas

Merle Wood
Education Consultant
Formerly of Oakland Public Schools
Lafayette, California

South-Western Publishing Co.

ISBN: 0-538-70768-2

2 3 4 5 6 7 8 9 0 DH 98 97 96 95 94 93

Printed in the United States of America

Developmental Editor: Gayle Entrup
Production Editor: Martha G. Conway
Associate Director/Design: Darren Wright
Production Artist: Steve McMahon
Associate Photo Editor/Stylist: Linda Ellis
Marketing Manager: Shelly Battenfield

PHOTO CREDITS

Page 23, left: Monroe Systems for Business

This book is printed on recycled, acid-free paper that meets Environmental Protection Agency standards.

PREFACE

Basic skills are required for each of us to conduct our personal and business dealings. These basic skills are vital to improving both our personal interactions and employment opportunities. *Basic Math With a Calculator* is specifically designed to provide adults with these basic skills.

STRUCTURE AND ORGANIZATION

Basic Math With a Calculator presents clear objectives followed by short segments of material with specific exercises for immediate reinforcement. This competency-based approach enables students to achieve success frequently as they progress through the material building their self-esteem and competency.

The organization of *Basic Math With a Calculator* reflects the emphasis on learning to use the calculator with basic math in an everyday setting. In *Basic Math With a Calculator,* students review the four basic functions of math. Students are evaluated after completion of the module.

The Pretest, Posttest, Glossary, Index, Answers, and Personal Progress Record are designed to facilitate and enhance independent student learning and successful achievement.

FEATURES OF *BASIC MATH WITH A CALCULATOR*

Basic Math With a Calculator is a complete and comprehensive package providing the student with learning material written specifically to meet the unique needs of the adult learner, and providing the instructor with support materials to facilitate student success. Some special features of the text-workbook include the following:

- *Appropriate Content.* Real-life issues and skills are emphasized throughout the text with relevant examples and illustrations to which the student can relate. A larger typeface is used to make the text-workbook easier for the student to use and to read.
- *Self-Contained Approach.* The text contains all the materials needed for self-paced and independent learning. A pretest and posttest allow students to evaluate their own skills. Answers for both tests and all exercises are also included.
- *Clear Objectives.* Instructional objectives for each unit are clearly stated, letting students know what they will learn.

- *Abundant Exercises with Goals.* Each unit provides specific exercises following short segments of instruction, allowing students to immediately use what they have learned. The end-of-unit activities provide additional applications for further reinforcement. All exercises provide goals for students to measure their own skill development and success, enhancing student motivation and direction.
- *Interesting Study Breaks.* Each unit contains study breaks that provide a refreshing change of pace and yet contribute to the general literacy goal of the student.
- *Positive Summaries.* A summary of the student's accomplishments is presented at the end of each unit, providing encouragement and reinforcement.
- *Personal Progress Record.* Students keep track of their own progress by recording scores on a Personal Progress Record. Students can measure their own success by comparing their scores to evaluation guides provided for each unit. Whenever a student's total score for a unit is below the minimum requirement, the student may request a Bonus Exercise from the instructor for further application.

FEATURES OF THE INSTRUCTOR'S MANUAL

The Instructor's Manual provides general instructional strategies and specific teaching suggestions for *Basic Math With a Calculator,* along with supplementary bonus exercises and answers, additional testing materials, and a certificate of completion.

Bonus Exercises

Second-chance exercises for all activities are offered through bonus exercises provided in the Instructor's Manual. These bonus exercises enable instructors to provide additional applications to those students whose scores are less than desirable for a unit. Answers to all bonus exercises are also provided and can be reproduced for student use.

Testing Materials

Two additional tests are provided in the manual to allow for more flexible instruction and evaluation.

Certificate of Completion

Upon completion of *Basic Math With a Calculator,* a student's success is recognized through a certificate of completion. This certificate lists the topics that were covered in this text-workbook. A certificate master is included in the manual.

CONTENTS

GETTING ACQUAINTED

Most of us can use a hand-held calculator to solve math problems. You often need to use the calculator to solve math problems at home, at work, and at school. You may need to use the calculator to figure how much you are saving when you use coupons at the grocery store or the number of hours you worked, or to add your grades in a course at school.

You may be enrolled in this math program because you know your math skills need improvement. Or you might be taking it to brush up on calculator skills. This text-workbook will help you to reach either objective.

WHAT'S GOOD ABOUT CALCULATORS?

You will like working with a calculator whether you are good at math or you need to improve your math skills. Everyone likes to work with calculators because it is easier, faster, and more accurate to press a key on a calculator than it is to add a number in your head. While you are doing math by calculator, you will also be learning how to solve math problems.

HOW YOU WILL LEARN

Basic Math With a Calculator is designed specifically with you in mind. You will learn quickly and easily about using a calculator to solve common basic math problems at your own pace.

Learn at Your Own Pace

You will progress through this text-workbook at your own pace. You may move ahead faster, or go slower, than other students. But don't be concerned about this. You are to work at *your* best speed.

Learn Skills Successfully

You will know what you are to accomplish. You will study a topic. Then you will drill over what you have learned. When you have shown that you know the topic, you will move on to the next topic. If you have not learned the topic, you will do added work on that section. You will know just how well you are doing as you move through each step in this book.

Complete Bonus Exercises

You may not reach your assigned goal on every practice activity. When this happens, you should review the lesson and then do a

Bonus Exercise. These bonus exercises cover the same material as the practice exercises in the book. They give you a second chance to reach your goal. Your instructor has copies of these bonus exercises and the answers to them.

Check Your Own Success

You will keep track of your own success. You will check all of your own work. The answers are in the back of the book. Then you will record your score on your own Personal Progress Record, which is also at the back of the book.

WHAT YOU WILL LEARN

As you study *Basic Math With a Calculator,* you will learn the basic skills needed to do calculator math, including:

- Find correct answers to addition, subtraction, multiplication, and division problems.
- Write numbers so they are easy to read.
- Proofread numbers, estimate answers, and prove answers.
- Solve word problems.
- Round numbers to tens, hundreds, and thousands places.
- Find correct answers to money problems.

SPECIAL FEATURES

Basic Math With a Calculator has a number of special features. These features will help you learn quickly and apply the material successfully.

Checking What You Know

You can check what you already know before starting this book. *Checking What You Know* lets you know what skills you need to improve upon. By comparing your score on that pretest with your score on *Checking What You Learned* at the end of the book, you will see how much you have gained through your study.

Break Times

The three break times will give you a rest from study. These *Calculators Make It Easy* sections show you how calculators simplify math problems. The *Just For Fun* sections show you how to play games on the calculator. The *Yesterday, Today, and Tomorrow* sections tell you some of the history of calculators.

Checkpoints and Activities

There are several exercises in each unit called *Checkpoints* and *Activities.* Each checkpoint will help you check your understanding of a specific topic before moving on to the next topic. At the end of each unit is a section called *Putting It Together* that contains several Activities. These are similar to the Checkpoints within the units. They will help you to apply and reinforce the skills you learned in the unit.

Bonus Exercises

If you do not reach the assigned goal for any one of the Checkpoints or Activities, you should review the unit. Then you will be asked to do a Bonus Exercise. This bonus work gives you a second chance to succeed. These second-chance activities are not in this book. Your instructor has copies of these bonus exercises for you, along with the answer key to these activities. You will use the answer key to check your own work.

Answers

Answers to all the exercises are provided at the back of this book. The color of the answer pages makes them easy to find and use. You will use these pages to check your own work. Always do the exercises *before* you look at the answers. Use the answers as a tool to check your work—not as a means of completing the exercise.

Personal Progress Record

You will check your own work with the answers. Then you will record your score on your own *Personal Progress Record* located at the back of this book. After you complete a unit, you will be able to determine your level of success.

Certificate of Completion

When you finish your study in this book you may be eligible for a certificate of completion. Your instructor will explain the skill level required for this award.

READY TO START!

You are now ready to start improving your basic math skill for personal and job use. Through your independent study and the completion of the exercises, you should develop improved ability in calculator math.

CHECKING WHAT YOU KNOW

Take this pretest before starting *Basic Math With a Calculator*. The 30 questions will help you determine what skills you already know well. They will also give you an idea of those skills you should improve upon. There is no time limit, so take the time you need.

Follow the directions for each section. Write your answers in the space provided. When you finish, check your work against the answers near the back of this book. Give yourself 1 point for each correct answer. Record your score in your Personal Progress Record. Use the analysis chart to see where you need to study to improve your skill.

DIRECTIONS: Use your calculator to find the answers to all of the problems below. Write your answers in the blank spaces.

(1) Find the total. 173 + 68 + 6,532 = **1.** ______

Column A shows the correct number. Draw a line through the number in Column B if it is different from the number in Column A. Write in the blank space the correct number from Column A.

	Column A	Column B	
(2)	$3,928.18	$3.7ZS.1S	**2.** ______

(3) Round 398 to the nearest hundreds place. **3.** ______

Work the following problem: 439 + 526

(4) Rounded Problem **4.** ______

(5) Estimated Total **5.** ______

(6) Calculator Total **6.** ______

Underline the clues to these addition problems. Then work the problems to find the correct totals.

(7) Jenny is in a weight-reducing class. In three months, she lost 13 pounds, 9 pounds, and 7 pounds. How many pounds did she lose all together? **7.** ______

(8) Allen wants to estimate the miles to his aunt's house. The map says the distance is 165 miles. Round these miles to hundreds. **8.** ______

(9) Find the total.

$$\begin{array}{r} \$13.64 \\ .70 \\ 21.49 \\ 11.87 \\ +\ 29.63 \\ \hline \end{array}$$

9. ______

(10) Find the difference. 235 − 168 = 10. ______

Estimate and calculate the differences for this problem: 189 − 58

(11) Rounded Problem 11. ______

(12) Estimated Difference 12. ______

(13) Calculator Difference 13. ______

Underline the clue that tells you to subtract in this word problem. Find the difference.

(14) Cindy's grocery bill total came to $24.58. She gave the cashier $25. How much money does she have left over? 14. ______

(15) Find the difference.

$$\begin{array}{r} \$328.06 \\ -\ 18.73 \\ \hline \end{array}$$

15. ______

(16) Find the product.

$$\begin{array}{r} 754 \\ \times\ 33 \\ \hline \end{array}$$

16. ______

(17) The community theater sold out all 7 nights of its performances. The theater seats 635 people. How many people saw the play? 17. ______

(18) Round 7.3946 to hundredths. 18. ______

Estimate and calculate the products for this problem: 526 × 89

(19) Rounded Problem 19. ______

(20) Estimated Product 20. ______

(21) Calculator Product 21. ______

(22) Find the quotient. Round to hundredths. 534 ÷ 73 = 22. ______

(23) Doyle made 4 trips to work and back home. He traveled 84 miles. How far is Doyle's trip to work and back home? 23. ______

Find the Unrounded Quotient. Then find the Rounded Quotient. Round to two decimal places. 572 ÷ 24

(24) Unrounded Quotient 24. ______

(25) Rounded Quotient 25. ______

Estimate and calculate the quotients for this problem. Round the calculator answer to two decimal places. 635 ÷ 49

(26) Rounded Problem **26.** ______

(27) Estimated Quotient **27.** ______

(28) Calculator Quotient **28.** ______

Find the quotient. Round to two decimal places.

(29) Amanda worked 38 hours last week. She got paid $278.48. How much does Amanda earn an hour? **29.** ______

Use chain division to solve this problem. Round your answer to two decimal places.

(30) 479 ÷ 26 ÷ 43 **30.** ______

Check your work on page 61. Record your score on page 65.

UNIT 1 Addition—the Basics

WHAT YOU WILL LEARN

When you finish this unit, you will be able to:

- Get the correct answers to addition problems.
- Write the numbers 0–9 so they are easy to read.
- Proofread numbers with different styles and sizes.

THE BASICS OF ADDITION

On the job and in your personal life, you will probably solve more addition problems than any other kind of math problem.

Addition means combining two or more numbers, called **addends,** to get a total. A **total** (also called a **sum**) is the answer to an addition problem. For example, if you were setting the supper table for three people in your family and suddenly two guests arrived, you would add two more place settings. The addends are 3 and 2. The total or sum is 5.

	Place Settings	
Your family:	3	Addend
Guests:	+ 2	Addend
	5	Total (or Sum)

Addends may have one or more digits. A **digit** is one of the numbers 0, 1, 2, 3, 4, 5, 6, 7, 8, or 9. When there are only a few addends with one digit each, it is easy to add the numbers in your head. For example, $3 + 2 = 5$ or $2 + 4 + 7 = 13$. When there are many numbers with many digits each, it is hard to add the numbers in your head. For example, $6{,}529 + 48 + 3{,}710$. For this problem, it will be easier and more accurate to use a calculator.

Your boss at the Pastry Shop asks you to find out how many pastry boxes are in the storeroom. You count them and find there are 58 large, 39 medium, and 61 small boxes. You need to add the three addends to get the total. Since it is hard to add those numbers in your head, you will use a calculator.

Calculator Addition

Your calculator may look like the one in Illustration 1-1.

Don't worry if your calculator is slightly different. It will still work the same as the one in the picture.

To add the addends 58, 39, and 61:

1. First, turn your calculator ON. Tap the [ON/C] On/Clear Key or [CE/C] Clear Entry/Clear Key. Some calculators are always ON and ready to use.
2. Clear any numbers out of the calculator. Tap the [AC] All Clear Key, or the [CE/C] Key.
3. Enter the addends by pressing these keys:

   58 + 39 + 61 = 158 ______

   The total should be in the answer window. Does your answer match the answer in the book (158)? If your answer is different from the book's, do the problem until you get the correct answer.
4. Write your answer on the blank line after the problem.
5. Clear the calculator again. Make it a habit to clear your calculator after each problem so the next problem starts with a 0 in the window.

Illustration 1-1

A hand-held calculator.

Vertical Addition

Sometimes addends will be listed **vertically** (up and down) instead of **horizontally** (from left to right). Enter the addends the same as you would for horizontal problems. For the following problem you would tap the keys 58 + 39 + 61 as you did before.

```
   58
   39
+  61
```

CHECKPOINT 1-1

YOUR GOAL: Get 4 of 5 answers correct.

Before you work these problems, find the number keys from 0 to 9, the [+] Plus Key, and the [=] Equal Key on your calculator. If your answers are not the same as the book's answers, do the problems until your answers are the same as the book's answers.

- ● 23 + 249 + 4 = **276**
1. 6 + 48 + 29 = ____
2. 7 + 39 + 648 = ____
3. 273 + 8 + 91 = ____
4. 47 + 483 + 9 = ____
5. 520 + 142 + 308 = ____

Check your answers on page 61. Then count the number of answers you got correct. Record your score on the Personal Progress Record on page 66.

CALCULATORS MAKE IT EASY!

Most of us like work that is easy to do. Addition problems can be hard to do in your head when you have to remember a number and then "carry" it to the next column to the left.

```
   ¹
   58
   37
+  61
  156
```

The 1 in 16 (the total of the right column) is mentally "carried" to the next column to the left.

You will like doing addition problems on the calculator because it will automatically "carry" the 1 over to the next column to the left.

BEING ACCURATE

Calculators help you get totals quickly—but it is more important to be correct than to be fast. A total you get quickly is no good if it is not correct. Here is what you can do to get correct totals:

See, Read, Remember, and Write Totals

To get a correct answer, you must take time to see and read each number correctly, remember the number in your head, and then enter that same number into the calculator.

Think about what you are doing. Read numbers in groups if they are separated by commas. Pause where there is a comma as in Illustration 1-2.

Illustration 1-2

See, read, remember, and write numbers.

	See, Read	*Remember*
a.	89	eight nine
b.	574	five seven four
c.	6,458	six [pause] four five eight

CHECKPOINT 1-2

YOUR GOAL: Get all answers correct.

Write in the blank spaces what you see, read, and remember.

	See and Read	Write
●	54	five four
1.	72	
2.	371	
3.	698	
4.	8,964	
5.	2,514	

Check your work on page 61. Record your score on page 66.

Write Numbers Legibly

Writing numbers **legibly** means writing them clearly. You must write numbers legibly so that you and others can read them quickly and correctly. To avoid mistakes, take the time to form your numbers carefully.

When numbers are not written legibly, it is hard to know what the numbers are. For example, look at Illustration 1-3. Write the

Illustration 1-3

Illegible handwritten numbers.

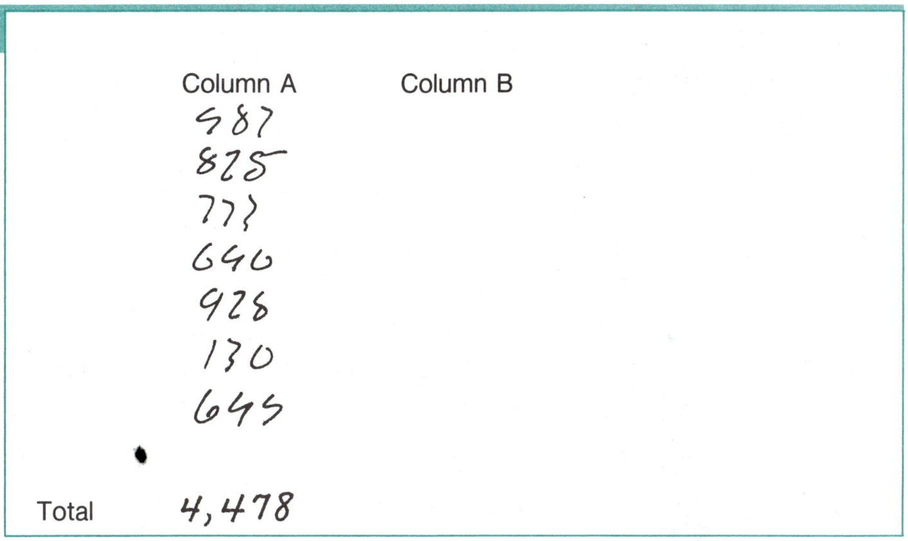

numbers that you believe are represented in Column A in Column B. Do your numbers add up to the same total?

CHECKPOINT 1-3

YOUR GOAL: Get a 3 rating for Step 3. The ratings are:

* = 1
** = 2
*** = 3
**** = 4

To help you write numbers legibly, do this task:

Step 1: Write the numbers 0 through 9 on the line below:

____

Step 2: Now compare your numbers with those in Illustration 1-4 on the 4-star (****) line of the Number Rating Scale.

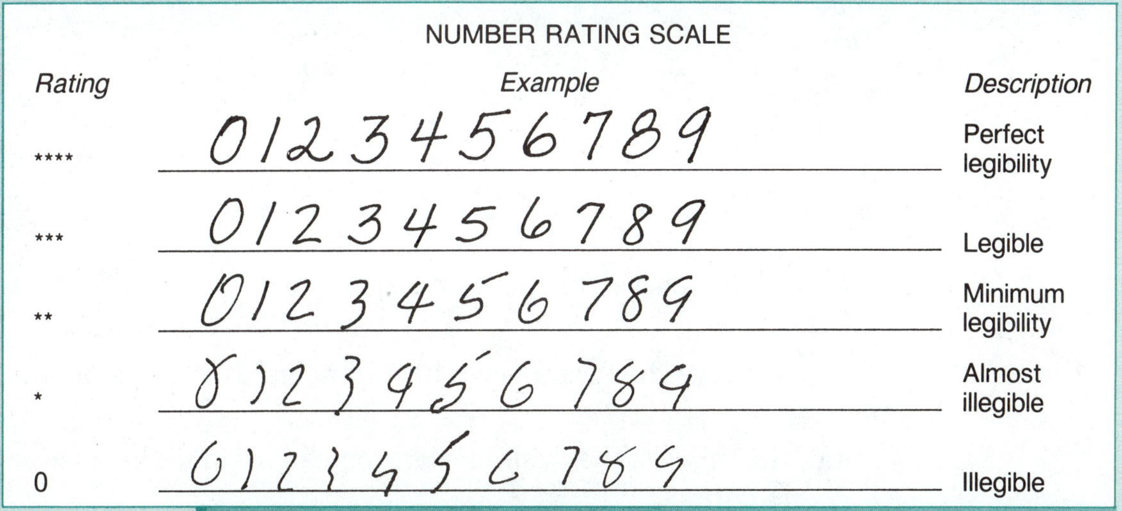

Illustration 1-4

Numbers rated from Perfect Legibility to Illegible.

(1) Are all the numbers the same size?
(2) Are the lines straight in the numbers 1, 4, and 7?
(3) Are the angles in the numbers 4, 5, and 7 straight and not rounded?
(4) Are the circles in the numbers 6, 8, 9, and 0 closed?
(5) Are the sizes of the numbers and the spaces between numbers written so they fit the length of the line on which they are written?

Step 3: Writing legible numbers requires attention and careful practice. Practice again writing legibly the numbers 0 through 9 until they have a 3-star or 4-star rating.

___

Check your work on page 61. Record your score on page 66.

Prove the Total

To **prove the total** for an addition problem means to do the problem a second time in a different way. The second time, add from the bottom to the top of the problem. The chance of you making the same mistake a second time is less if you add them in a different order. For example, if you first add the numbers from bottom to top, next time add from top to bottom. Or from the right to the left instead of from the left to the right. To prove a total:

Step 1: Clear your calculator.

Step 2: Work the problem by adding down:

Add
Down

```
   58  |
   39  |
 + 61  ↓
 ----
```

Record the answer in the blank space.

Step 3: Prove the first answer by adding up. If the second answer is:

(a) the *same* as the first answer, you can assume the answer is correct.
(b) *different* from the first answer, repeat the problem until you get two answers that match. Draw a line through the incorrect answer.

Add
Up

```
   58  ↑
   39  |
 + 61  |
```

Record the answer in the blank space.

Step 4: For horizontal problems, add from left to right:

→
58 + 39 + 61 = _____

and then from right to left to prove the total:

←
58 + 39 + 61 = _____

Proofread Numbers

Proofreading numbers means examining (looking at) digits in a number to see if they are the same or different. Numbers may appear in different styles and sizes, as shown in Illustration 1-5.

Illustration 1-5

Proofread carefully.

	Column A	Column B
1.	04371	04317
2.	32,413.88	$2,431.88
3.	471983	471938
4.	May 13, 1987	May 13, 1987
5.	Pages 324-491	Pages 342-419

CHECKPOINT 1-4

YOUR GOAL: Get 4 of 5 answers correct for Column C only.

Draw a line through any number in Column B that is different from the number to be matched in Column A. Write the correct number in Column C. If the numbers in Columns A and B are the same, write the number in Column C. In Column D, rate your writing with stars after comparing each item with the examples in the Number Rating Scale, on page 6.

	Column A	*Column B*	*Column C*	*Column D*
●	72HB609	~~72HB909~~	72HB609	* * * *
1.	6431 ÷ 839	6431 ÷ 839		
2.	11:03 a.m.	11:03 a.m.		

3.	214-401-5970	214-481-4970	________	________
4.	8/14/88	8/14/88	________	________
5.	368 907	368 907	________	________

Check your work on page 61. Record your score on page 66.

WHAT YOU HAVE LEARNED

Now that you have completed this unit, you should be able to:

- Correctly add numbers.
- Write numbers legibly.
- Proofread typewritten and handwritten numbers.

PUTTING IT TOGETHER

ACTIVITY 1-1 YOUR GOAL: Get all answers correct.

Write your answers in the blank spaces.

What is the total?

●	274 +	6 +	19			=	299
1.	649 +	382 +	27 +	570		=	______
2.	728 +	385 +	482 +	42		=	______
3.	2,819 +	349 +	5,791			=	______
4.	315 +	827 +	98,630			=	______
5.	128 +	46 +	4,318 +	24,891		=	______

Check your work on page 61. Record your score on page 66.

ACTIVITY 1-2 YOUR GOAL: Get all answers correct.

The correct numbers are in Column A. Draw a line through any number in Column B that is different from the number to be matched in Column A. Write in Column C the correct numbers from Column A.

	Column A	Column B	Column C
●	61897	~~61391~~	61897
1.	**$1,892.43**	$1,982.43	______
2.	12:38 p.m.	12:08 p.m.	______
3.	*SQ-178-3062*	SO-718-3062	______
4.	9/24/93	9/24/96	______
5.	**15973628**	160180ZS	______

Check your work on page 61. Record your score on page 66.

UNIT 2 Addition—Using It Every Day

WHAT YOU WILL LEARN

When you finish this unit, you will be able to:

- Round numbers to tens, hundreds, and thousands places.
- Estimate the answers to addition problems.
- Use a six-step process to solve word problems.
- Get the correct answers to addition, decimal, and money problems.

ROUNDING NUMBERS

To **round** a number means to change a number to the nearest power of ten such as tens (10), hundreds (100), or thousands (1,000). Rounded numbers make it easier to quickly understand amounts. For example, it is easier to "grasp in your mind" $10 instead of $9.95. Each digit in a number has a value depending on its place in a number. Illustration 2-1 shows the place value of digits in the number 7,146.

Illustration 2-1

Place value chart.

Places

Thousands	Hundreds	Tens	Ones
7	1	4	6

The 6 is in the *ones* place.
The 4 is in the *tens* place.
The 1 is in the *hundreds* place.
The 7 is in the *thousands* place.

To round 7,146 to the nearest *hundred*:

1. Look at the digit in the place you are rounding and underline it. (Underline 1 because 1 is in the *hundreds* place and we are rounding to the nearest hundred.)
2. Look at the digit to the right of the place you are rounding and circle it. (Circle 4 because it is to the right of 1.)
3. If the digit is
   a. 5 or *more,* add 1 to the place you are rounding. (4 is *not more than 5,* so go to step b.)
   b. *less* than 5, keep the same number in the place you are rounding. (4 is *less than 5,* so keep the 1.)
4. Change all digits to the right of the place you are rounding (1) to zeros. (Change the 4 to 0 and the 6 to 0 so 146 changes to 100. 7,146 is rounded to 7,100, the nearest hundred.)

Now round 7,146 to the nearest *ten* by following the steps below and Illustration 2-2:

1. The rounding number is 4 in the *tens* place. Underline it.
2. The number to the right of 4 is 6. Circle it.
3. 6 is more than 5. So add 1 to the 4, the place you are rounding, which changes the 4 to 5.
4. Change the 6 to 0. 7,146 rounded to the nearest ten becomes 7,150.

Illustration 2-2

Rounding steps.

Step

1. 7,146
2. 7,146
3. 7,15__
4. 7,150

CHECKPOINT 2-1

YOUR GOAL: Get 4 of 5 answers correct.

Round each number to the place shown.

● 2,476 to the nearest tens place. **2,480**

1. 326 to the nearest tens place. ______

2. 5,078 to the nearest hundreds place. ________

3. 3,479 to the nearest thousands place. ________

4. 7,551 to the nearest hundreds place. ________

5. 4,623 to the nearest thousands place. ________

Check your work on page 61. Record your score on page 66.

ESTIMATING TOTALS

"Accuracy first" is our motto for all math problems. You already know some ways to get the correct answer from the Being Accurate section. Another accuracy check is to **estimate** (or judge) the total.

To estimate the total of 47 + 631:

1. Round each addend to the highest place at the left. (47 becomes 50 and 631 becomes 600.) Write the rounded addends in the Rounded Addends column.
2. Mentally add the Rounded Addends and write the total 650 in the Estimated Total column.
3. Use the calculator to add the original addends. (47 + 631). Write the total in the Calculator Total column (678).
4. If the Calculator Total is not close to the Estimated Total, estimate and calculate the answer again. Write the correct Estimated Total and Calculator Total.

Original Addends	**Rounded Addends**	**Estimated Total**	**Calculator Total**
47 + 631	50 + 600	650	678

CHECKPOINT 2-2

YOUR GOAL: Get 12 of 15 answers correct.

Work the following problems.

	Original Addends	Rounded Addends	Estimated Total	Calculator Total
●	691 + 178	700 + 200	900	869
1.	17 + 41	________	________	________
2.	34 + 740	________	________	________
3.	271 + 426	________	________	________

4. 1,729 + 5,493 ________ ________ ________

5. 1,830 + 46,925 + 7,026 ________ ________ ________

Check your work on page 61. Record your score on page 66.

SOLVING WORD PROBLEMS—ADDITION

In textbooks, math problems are often written as number problems (with no words) such as:

$$\begin{array}{r} 56 \\ 39 \\ +\ 61 \\ \hline \end{array}$$

In real life, math problems are not "set up" for you but are part of a "story" that you must study to get the answer. These problems are sometimes called **word problems.** For example:

Your boss asks you to find the *total* number of boxes in the storeroom. You count the boxes and find out that there are 58 large, 39 medium, and 61 small boxes.

Problem-Solving Process

There is an easy six-step process that will help you solve math word problems.

Step	To Do	Explanation	Example
1	Read the story	Look for and underline key words.	*total number of boxes, 58 large, 39 medium, 61 small*
2	Answer key questions	What information is:	
		a. Needed?	a. *Total number of boxes*
		b. In the story?	b. *58, 39, 61*
3	Decide which math operation to use	+, −, ×, or ÷	The story says to "total" so you must add.
4	Estimate the answer	About what should the answer be?	The answer will be about 150 since all of the addends are about 50 and mentally you know that three 50's = 150.
5	Work the problem	Calculate accurately.	58 + 39 + 61 = 158
6	Prove the answer	Calculate the addends in reverse order.	61 + 39 + 58 = 158

Clues to Look For

You should use addition when words in a problem mean the same as *add.* For example, in the following story "put together" means the same as *add,* so this is an addition problem.

James and Jenny want to buy a Mother's Day gift. James has \$6 and Jenny has \$8. How much can they spend on a gift if their money is put together?

Some examples of words that mean the same as *add* are:

accumulate
all together
combine or combines or combined
in all
join or joins or joined
put together
sum
total

These words (and other words like them) are the clues that will tell you if the story is an addition problem.

CHECKPOINT 2-3

YOUR GOAL: Get 4 of 5 answers correct.

Read the following stories and find the "clues" that will tell you if it is an addition problem. Underline the clue.

If it is an addition problem, get the answer. If it is not an addition problem, write "No +" in the answer space.

- You work at Bigger Burgers. The supervisor asks you to take inventory of the cups in the stockroom and front serving area. You count 165 cups in the stockroom and 59 cups in the front serving area. How many cups are there in all?

165 + 59 = **224**

1. Pauline wants to get a hair dryer at the trading stamp store with her stamps. Her stamp books are in three stacks; 8 in one stack, 9 in another stack, and 16 in another stack. How many books does she have all together?

____

2. Mary has \$15 in her purse. She finds a pair of shoes on sale for \$10 and decides to buy them. How much money does she have left in her purse?

____

3. Carl and Sylvia decided to shop for a new car. They saw one they liked. A sheet of paper taped to the car window showed the price of the car and other charges:

Base Price	$5,995
AM/FM Stereo Radio	235
Automatic Transmission	180
Air Conditioning	574
Destination Charge	246

What is the total price of the car? ______

4. George and Fred were hired to paint a garage for Ms. Bush. Ms. Bush paid them $200. How much did George make after the money was divided evenly between the two men?

______

5. Donna was shopping at a second-hand store. She decided to buy a purse for $8, a pair of shoes for $13, and a pair of sandals for $9. How much did she spend at the store?

______

☞ *Check your work on page 61. Record your score on page 66.*

DECIMALS AND AMOUNTS OF MONEY IN ADDITION

The [·] Decimal Point Key is pressed at the place in the addend where the decimal occurs. To enter 3.81 + 0.75 in the calculator, press 3 [·] 81 + [·] 75 = . The total is 4.56.

Amounts of money are written as dollars and cents with two decimal places. To get the total of a grocery purchase, tap the numbers and decimal keys as shown below:

$$\begin{array}{r} \$\ 3.81 \\ 2.25 \\ 1.02 \\ +\quad .79 \\ \hline \$\ 7.87 \end{array}$$ ______

Record your answer on the blank line.

DECIMALS NOT ALIGNED

Sometimes decimals will not be **aligned** (stacked in a straight line below each other). Just enter the decimal point at its proper place in each addend. The answer will appear in the window with the decimal accurately placed, as shown in Illustration 2-3.

```
  1.736
  5.98
+ 2.4
```

Illustration 2-3

Your calculator automatically aligns the decimal.

CHECKPOINT 2-4

YOUR GOAL: Get 4 of 5 answers correct.

Find the totals.

```
•     21
      3.78
  + 475.95
    500.73
```

```
1.   15.80     2.  $42.39    3.  $ 92.37    4.   5.633    5.     .67
     23.67           3.67          8.98          7.89          43.27
   +  4.05       +  55.55     +  639.24      + 138.4       + 1.7849
```

Check your work on page 61. Record your score on page 66.

CALCULATORS MAKE IT EASY!

We like to use calculators—they make our work easier than doing the same work by hand or in our head. For example, to add numbers with decimals without using a calculator, take the time to rewrite all of the addends with unaligned decimals in Column A so the decimals will be aligned in Column B:

Column A	*Column B*
Unaligned decimals	Addends rewritten to align decimals
6.049	
26.1	
7.37	
+ 857.3	______
	______

Now, mentally add the numbers you just rewrote in Column B and record the answer.

To compare how easy it is to use a calculator for addends with decimals, add the numbers in Column A with your calculator.

WHAT YOU HAVE LEARNED

Now that you have completed this unit, you should be able to:

- Round numbers.
- Estimate answers for addition problems.
- Correctly add numbers with decimals and amounts of money.
- Solve word problems.

PUTTING IT TOGETHER

ACTIVITY 2-1 YOUR GOAL: Get 4 of 5 answers correct.

Round each number to the place suggested.

- 5,291 to the nearest thousands place. 5,000
1. 29 to the nearest tens place. ______
2. 83 to the nearest tens place. ______
3. 758 to the nearest hundreds place. ______
4. 3,649 to the nearest thousands place. ______
5. 9,148 to the nearest thousands place. ______

Check your work on page 61. Record your score on page 66.

ACTIVITY 2-2 YOUR GOAL: Get 12 of 15 answers correct.

Work the following problems.

	Original Addends	Rounded Addends	Estimated Total	Calculator Total
●	871 + 392	900 + 400	1300	1263
1.	35 + 54			
2.	47 + 197			
3.	639 + 216			
4.	799 + 924			
5.	108 + 593			

Check your work on page 61. Record your score on page 66.

ACTIVITY 2-3 YOUR GOAL: Get 4 of 5 answers correct.

Where necessary, underline the clues to these addition problems. Then work the problems to find the correct totals.

- Jamie is in a weight-reducing class. In four months she lost 14 pounds, 12 pounds, 7 pounds, and 4 pounds. How many pounds did she lose all together?

  37

1. Your family drove to another state to visit relatives. The first day you traveled 328 miles, the second day 297 miles, and the last day 405 miles. In all, how many miles did your family drive?

______

2. Rob wants to estimate the miles to his aunt's house. The map says the distance is 383 miles. Round these miles to hundreds.

______

3. Janet is planning to buy a pair of shoes for about $39, a blouse for about $22, and a pair of pants for about $38. About how much money will she need all together?

______

4. Kaye worked in a grocery store 37 hours the first week of August, 29 hours the second week, 40 hours the third week, and 33 hours the fourth week. How many hours did she work total?

______

5. Mandy went to the grocery store with very little money. Before checking out, she wants to make sure she has not put too many groceries into her cart. Below is a list of prices of the items she had in her grocery cart. What was the total of the grocery items?

$$\begin{array}{r} \$0.39 \\ 0.69 \\ 1.19 \\ 0.89 \\ +\ \ 0.24 \\ \hline \end{array}$$

______

Check your work on page 62. Record your score on page 66.

ACTIVITY 2-4 YOUR GOAL: Get 4 of 5 answers correct.

Write the totals in the blank spaces.

• $$\begin{array}{r} 58.53 \\ 2.31 \\ .20 \\ 9.32 \\ +\ 25.80 \\ \hline 96.16 \end{array}$$

1.	2.	3.	4.	5.
25.63	$36.45	$19.28	793.3	9.263
.78	2.98	26.43	2.86	43.8
9.41	.37	.29	5.941	.57
43.07	.03	1.56	.3709	.453
+ .56	+ 19.85	+ .75	+ 520.8	+ 673.1

Check your work on page 62. Record your score on page 66.

UNIT 3

Subtraction—Using It Every Day

WHAT YOU WILL LEARN

When you finish this unit, you will be able to:

- Get the correct answers to subtraction problems.
- Prove the answers to subtraction problems.
- Estimate the answers to subtraction problems.
- Find clues to subtraction problems.
- Get the correct answers to subtraction of decimals, amounts of money, and series of subtrahends problems.

THE BASICS OF SUBTRACTION

Subtraction is taking one number away from another number. The number *being subtracted* (the lower number) is called the **subtrahend,** and the number *being subtracted from* (the upper number) is called the **minuend.** The answer to a subtraction problem is called the **difference.**

59	Minuend
− 46	Subtrahend
13	Difference

Calculator Subtraction

Suppose the starter on your car does not work. You need to buy a new one that costs $43 so you can get to work. You now have only $16. How much more money do you need to buy the starter? You need to subtract the amount of money you have from the price of the starter to find how much more money you need. This is difficult to do in your head, so you will use a calculator.

To subtract 16 from 43:

1. Make sure your calculator is on.
2. Clear the calculator by pressing the [AC] key, or the [CE/C] key.

3. Enter the minuend, 43.
4. Locate the [–] Minus Key on your calculator as shown in Illustration 1-1 on page 000. Press the [–] Minus Key.
5. Enter the subtrahend, 16.
6. Press the [=] Equal Key.

The answer, 27, should be in the answer window. If it is not, work the problem until you get the correct answer.

Horizontal Subtraction

Minuends and subtrahends can be written vertically or horizontally, such as: 193 – 52 = 141. For this problem, press the keys **193 – 52 =**. Subtraction problems written horizontally are worked the same way as problems written vertically.

PROVE THE DIFFERENCE

To prove answers in subtraction problems, add the difference to the subtrahend. If the total is equal to the minuend, your answer is correct. In other words, difference + subtrahend = minuend. For example:

59	Minuend	Proof:	46	Difference
– 13	Subtrahend		+ 13	Subtrahend
46	Difference		59	Minuend

CHECKPOINT 3-1

YOUR GOAL: Get all answers correct.

To practice solving subtraction problems listed vertically and horizontally, work the problems below. Prove your answers by adding the difference and the subtrahend to get the minuend. Write your answers in the blank spaces.

● 87	**Proof:**	56	Difference
– 31		+ 31	Subtrahend
56		87	Minuend

1. 35 − 14 = ____

2. 79 − 34 = ____

3. 343 − 89 = ____

4. 562 – 231 = ______

5. 7,430 – 856 = ______

Check your work on page 62. Record your score on page 66.

YESTERDAY, TODAY, AND TOMORROW

The first electronic calculator was introduced about 1970. It cost about $1,200. It was as large as a typewriter. One of the latest calculators, like yours, is about the size of a business card. Some small calculators cost about $4 to $10 and calculate faster than the early $1,200 calculators.

1980 Model

1990's Model

ESTIMATING DIFFERENCES

We estimate the difference of a subtraction problem to help us find errors. Estimating the answer of a subtraction problem is similar to the process of estimating the answer of an addition problem.

To estimate the difference for the problem 519 − 98:

1. In your head, round the minuend and the subtrahend to the highest place at the left. (519 becomes 500 and 98 becomes 100.) Write the rounded numbers in the Rounded Problem column.
2. Mentally subtract the rounded subtrahend from the rounded minuend and write the difference (400) in the Estimated Difference column.
3. Use the calculator to work the original problem (519 − 98). Write the difference in the Calculator Difference column (421).
4. If the Calculator Difference is not close to the Estimated Difference, estimate and calculate the difference again. Write the correct Estimated Difference and Calculator Difference.

Original Problem	*Rounded Problem*	*Estimated Difference*	*Calculator Difference*
519 − 98	500 − 100	400	421

CHECKPOINT 3-2

YOUR GOAL: Get 12 of 15 answers correct.

Estimate and calculate the differences to these problems and write your answers in the blank spaces..

	Original Problem	Rounded Problem	Estimated Difference	Calculator Difference
●	52 − 18	50 − 20	30	34
1.	83 − 44	______	______	______
2.	865 − 79	______	______	______
3.	184 − 92	______	______	______
4.	370 − 165	______	______	______
5.	276 − 138	______	______	______

Check your work on page 62. Record your score on page 66.

SOLVING WORD PROBLEMS—SUBTRACTION

You should use subtraction when the words in a problem mean the same as *subtract.* Some examples of words that mean the same as *subtract* are:

decrease or decreases or decreased
difference
left or left over
less or lessened
minus
reduce or reduces or reduced
remain or remains or remained
take away or took away

These words plus others like them are clues that let you know when to subtract. For example, *left over* is the clue in the following story that tells you to subtract.

Kristy was sick. The doctor gave her a bottle with 21 pills in it to cure her cold. The doctor said to take three pills a day for seven days. After five days, she had taken 15 pills. How many pills were left over in the bottle?

6

CHECKPOINT 3-3

YOUR GOAL: Get 4 of 5 answers correct.

Read the following stories and underline the clues that tell you it is a subtraction problem. Work the problems and write your answers in the blank spaces.

- In Julio's night class at the Adult Learning Center, there were 17 students the first day. Julio thought it would be great if he talked to some of his friends and they decided to come. About three days later, Julio counted 23 students in his class. What is the difference in the number of students on the first day of class and three days later?

  6

1. Jack needed to buy a used car. He found a car that got 27 miles to the gallon in town. He compared this to his friend's car that gets 19 miles to the gallon in town. What is the difference in miles per gallon between the two cars?

   ______

2. David weighed 220 pounds. He wants to lose weight so he will feel and look better. He went on a diet and now weighs 26 pounds less. How much does he weigh now?

   ______

3. Cindy went to the store to buy a can of corn for 69 cents. She has a coupon that reduces the price by 13 cents. How much will she pay for the can of corn?

   ______

4. You borrowed $26 to pay off an electric bill. You were soon able to pay back $17. How much money remains to be paid back?

   ______

5. Chelsea decided to make brownies for her son's birthday party. She made 24 brownies. The children at the party ate 17 of the brownies. How many were left for her family to munch on?

   ______

Check your work on page 62. Record your score on page 66.

DECIMALS AND AMOUNTS OF MONEY

Decimals in subtraction are used the same way as decimals in addition. Sometimes the decimal points may not be aligned in a problem. Enter the decimal point at its proper place in each number. The answer will be in the window with the decimal in the right place.

CHECKPOINT 3-4

YOUR GOAL: Get 4 of 5 answers correct.

Find the differences and write them in the blank spaces.

●	345.92
−	68.10
	277.82

1.	2.	3.	4.	5.
59.95	25.638	10.46	1634.7	193.65
− 3.65	− .5326	− .299	− 1.329	− 49.47

☞ *Check your work on page 62. Record your score on page 66.*

As with addition, amounts of money are written as dollars and cents with two decimal places. One number in a money problem may have a decimal ($39.64) and another may not, ($23). Enter numbers and decimals exactly as they are read. For example: **$39.64 − $23** = ______. Press the [=] Equal Key. The calculator will then place the decimal in its correct place in the difference. Record your answer in the blank space. The difference should be $16.64.

You may need to make subtractions when using coupons at a grocery store. This is easy to do. Just press the [−] Minus Key and the next subtrahend. Continue to do this until you have entered all of the subtrahends. Then press the [=] Equal Key. Enter the minuend and the subtrahends:

$31.28 − $.25 − $.75 − $.10 = ______

Record your answer in the blank space. Did you get $30.18?

The series of subtrahends may be listed vertically:

	72.23
−	.432
−	69.1
−	.29

The difference in the window of your calculator should be 2.408. Use the same process for subtracting several subtrahends in money problems. Here is an example:

$102.06 − $53.72 − $34.19 − $7.48 = ______

The difference in this problem should be $6.67.

CHECKPOINT 3-5

YOUR GOAL: Get 4 of 5 answers correct.

Work these problems and write your answers in the blank spaces.

- $22.57 − $.15 − $.33 − $.25 = **$21.84**

1. 10.6 − 5.23 = ____

2. $98.46 − 73.51 = ____

3. 10.95 − 1.489 − .352 = ____

4. $19.28 − $7.46 − $.57 − $.69 = ____

5. 54.7 − .384 − 1.539 = ____

Check your work on page 62. Record your score on page 66.

JUST FOR FUN

Find the answers to the following questions with your calculator. Work the problem and then turn your calculator upside down to read the answer.

1. What does a fish breathe with?

   65,218 − 4,930 − 1,929 − 643 = ____

2. Sometimes you get one in your cake batter.

   88,345,776 − 11,000,113 = ____

3. What is a round ball that shows all of the land and water on the earth?

   1,258,746 − 1,220,670 = ____

4. A snake's voice sounds like this.

   5,612 − 98 = ____

5. What do babies wear when they eat?

   1,120 − 302 = ____

WHAT YOU HAVE LEARNED

By completing this unit, you can now:

- Subtract numbers on the calculator with total accuracy.
- Prove differences.

- Estimate subtraction problem answers.
- Find clues that let you know when a word problem is a subtraction problem.
- Subtract decimals, amounts of money, and series of subtrahends.

PUTTING IT TOGETHER

ACTIVITY 3-1 **YOUR GOAL:** Get all answers correct.

Write your answers in the blank spaces. Prove your answers. What is the total?

●
$$\begin{array}{r} 250 \\ -\ 36 \\ \hline 214 \end{array}$$

1. $\begin{array}{r} 89 \\ -\ 54 \\ \hline \end{array}$

2. 455 − 96 = ______

3. 148 − 73 = ______

4. $\begin{array}{r} 675 \\ -\ 36 \\ \hline \end{array}$

5. $\begin{array}{r} 3{,}576 \\ -\ 245 \\ \hline \end{array}$

☞ *Check your work on page 62. Record your score on page 66.*

ACTIVITY 3-2 **YOUR GOAL:** Get 12 of 15 answers correct.

Estimate and calculate the differences for these problems and write your answers in the blank spaces.

	Original Problem	Rounded Problem	Estimated Difference	Calculator Difference
●	56 − 19	60 − 20	40	37
1.	52 − 21	______	______	______
2.	236 − 49	______	______	______
3.	464 − 92	______	______	______
4.	179 − 57	______	______	______
5.	382 − 264	______	______	______

☞ *Check your work on page 62. Record your score on page 66.*

ACTIVITY 3-3 YOUR GOAL: Get 4 of 5 answers correct.

Underline the clue that tells you to subtract in a word problem. Find the difference.

- The total of Miss Johnson's grocery bill is $13.92. She gave the cashier $15.00. How much money does she have left?

$1.08

1. Jacob's rent for his apartment increased from $189 to $245. What is the difference between the two amounts? ______

2. Cindy bought a head of lettuce for $0.73. When she handed the cashier a $1 bill, how much change did she get back? ______

3. Andrea had $200 in her budget for a new washing machine. Her old one won't work and can't be repaired. She found a machine at a garage sale for $125. How much less did she spend than she thought she would? ______

4. Mary's grocery bill total came to $12.36. She gave the cashier $15. How much money does she have left over? ______

5. On October 8, Annie Moore's balance in her savings account was $192.31. She withdrew $23. How much money remains in her account? ______

Check your work on page 62. Record your score on page 66.

ACTIVITY 3-4 YOUR GOAL: Get 4 of 5 answers correct.

Find the differences and write your answers in the blank spaces.

- 4.75 − .39 = **4.36**

1. 26.931 − 7.48
2. $357.45 − 18.90
3. 59,386.1 − .427
4. $94.75 − 2.81 − 30.97
5. 360.4 − 19.28 − .745

Check your work on page 62. Record your score on page 66.

REVIEW 3-1 YOUR GOAL: Get 5 of 7 answers correct.

Clothing on Sale

Mr. and Mrs. Mattison went to a summer sale at Wearwell's clothing store to buy clothes for their children. They want to find out how much they saved by going to the sale. Below is a list of items that they bought, the sale price, and the regular price. Follow these steps to find out how much they saved on each item.

1. Subtract the Regular Price from the Sale Price and write the answer in the Difference column.
2. Repeat step 1 for the other items.

	Regular Price	Sale Price	Difference
Jacket	$42.38	$34.98	$7.40
Pants	$37.99	$19.95	______
Socks	$7.00	$4.19	______
Shirt	$18.99	$14.25	______
Sweater	$27.99	$22.49	______
TOTALS	______	______	______

Find out how much they saved all together.

1. Add the Regular Price column and write the total in the blank space.
2. Add the Sale Price column and write the total in the blank space.
3. Subtract the Sale Price column from the Regular Price column and write the difference in the blank space.

Check your Difference total by adding the Difference column. Check your work on page 62. Record your score on page 66.

REVIEW 3-2 **YOUR GOAL:** Get 6 of 7 answers correct.

Coupons in Supermarket

Brenda went to the supermarket to buy groceries with coupons she cut out of the newspaper. Follow the steps below to find the amount she saved on each item.

1. Subtract the Less Coupon column from the Regular Price column.
2. Write the difference in the Cost with Coupon column.

Item	*Regular Price*	*Less Coupon*	*Cost with Coupon*
Cleanser	$1.89	$0.39	$1.50
Bread	$1.69	$0.50	______
Dog Food	$9.49	$2.25	______
Milk	$2.89	$0.49	______
Ground Beef	$2.13	$0.25	______
TOTALS	______	______	______

How much money did she save all together?

1. Add the Regular Price column and write your answer in the blank space.
2. Add the Less Coupon column and write your answer in the blank space.
3. Subtract the Less Coupon total from the Regular Price total and write your answer in the blank space.

☞ ***Check your answer by adding the Cost with Coupon column. Check your work on page 63. Record your score on page 66.***

UNIT 4 Multiplication—Using It Every Day

WHAT YOU WILL LEARN

When you finish this unit, you will be able to:

- Get correct answers to multiplication problems.
- Use clues to find multiplication word problems.
- Estimate answers to multiplication problems.
- Get correct answers to multiplication money problems.
- Round to get answers to multiplication problems with decimals.

THE BASICS OF MULTIPLICATION

This is a multiplication problem:

125	Multiplicand	(also called a factor)
× 3	Multiplier	(also called a factor)
375	Product	

The **multiplicand,** the number to be multiplied (125), and the **multiplier,** the number of *times* to multiply (3), are both also called **factors.** The 125 and the 3 are factors in this problem.

The **product** is the answer to a multiplication problem (375).

Calculator Multiplication

To solve the problem $125 \times 3 = 375$:

1. Enter the multiplicand, 125.
2. Press the [×] Multiplication Key.
3. Enter the multiplier, 3.
4. Press the [=] Equal Key.
5. The product, 375, should be in the answer window.

   If you did not get 375, do the problem again. Read each step slowly. Work the problem until you get 375.

6. Clear the calculator.

Prove the Product

Factors can be entered into the calculator in any order. The product will be the same (125 × 3 = 375, or 3 × 125 = 375).

To prove a multiplication product, multiply the multiplier by the multiplicand. For example:

125	Multiplicand	3	Multiplier
× 3	Multiplier	× 125	Multiplicand
375	Product	375	Product

Horizontal Multiplication

Sometimes a multiplication problem will be written horizontally (from left to right) instead of vertically (up and down). Enter the factors the same as you would for vertical problems.

CHECKPOINT 4-1

YOUR GOAL: Get 4 of 5 answers correct.

Find the products. Prove your answers. Write them in the blank spaces.

● 39 × 9 = **351** Proof: 9 × 39 = **351**

1. 42 × 7 = ______
2. 806 × 74 = ______
3. 1,643 × 835 = ______
4. 95 × 31 = ______
5. 581 × 392 = ______

Check your work on page 63. Record your score on page 67.

SOLVING WORD PROBLEMS—MULTIPLICATION

Multiplication is repeated addition. For example, 4 × 3 means 4 + 4 + 4. Four is added three times.

The clue to a multiplication problem is found by looking for a number that tells you to repeat another number a certain number of *times*.

For example, you and two friends decide to rent an apartment. Each of you must pay $120 per month. How much is rent per month?

Rent each Person	Person
120	1
120	2
+ 120	3
360	

So 120 added together three times equals 360. A shortcut for repeating addition is to multiply a number (120) a certain number of times (3). $120 \times 3 = 360$.

CHECKPOINT 4-2

YOUR GOAL: Get 4 of 5 answers correct.

Read the following problems. Find the clues that will tell you if it is a multiplication problem.

If it is a multiplication problem, write the answer in the blank space. If it is not a multiplication problem, write "No ×" in the answer space.

● Every cleaning worker at Hotel Haven cleans 25 rooms a day. Fourteen workers are on duty each day. How many rooms are cleaned each day?

350

1. You work at Ned's Nursery. The boss tells you to find out how many oak trees are in the nursery. There are five oak trees in each bunch and there are 27 bunches. How many oak trees are there all together?

____

2. Jim has $17 in his wallet. Paul has $5 in his wallet. How much money do the two have together?

____

3. Dave is a parking lot attendant. He cannot admit more cars than can be parked in his lot. Dave counts 45 parking spaces on one row. Each row has the same number of spaces. There are nine rows. How many cars can be parked in Dave's lot?

____

4. Kindra's telephone bill is $34. She sends the telephone company $23. How much does Kindra owe the telephone company?

____

5. Anthony jogs four miles every day. How far will he jog in 31 days? ______

Check your work on page 63. Record your score on page 67.

CALCULATORS MAKE IT EASY!

The solar energy that comes into the solar window of your calculator powers the calculator to automatically calculate multiplication problems.

For example, if you worked the following multiplication problem on paper, you would have to do three extra steps to solve it. If you work it on your calculator, you only have to press the correct keys to get the same answer. It also takes less time to press keys than to write numbers on paper.

By Pencil and Paper

7,543	Write the numbers
× 691	on paper.
7,543	You do the work
178,870	in your head and
4,525,800	by hand.
5,212,213	

By Calculator

7,543	Press the numbers
× 691	on a calculator.
	The calculator
	does the work
	automatically.
5,212,213	

ROUNDING UNROUNDED PRODUCTS

Products with decimals can be unrounded. This means the answer will have as many decimal places as there are in the multiplicand plus those in the multiplier.

For example:

1. Enter the problem.

		Decimal Places
Multiplicand	3.65	2
Multiplier	× 2.9	+ 1
Product	10.585	3

2. The product, 10.585, is an unrounded product (10.585 has three decimal places).

An unrounded product may be more exact than you need.

To round 10.585 to two decimal places, use the method on page 38.

The rounded product is 10.59.

CHECKPOINT 4-3

YOUR GOAL: Get 8 of 10 answers correct.

Find the Unrounded Product. Round each number to the number of places in the Round To column.

		Round To	Unrounded Product	Rounded Product
●	3.28 × 57.6	2 places	188.928	188.93
1.	7.43 × 8.02	2 places	______	______
2.	9.138 × 0.647	3 places	______	______
3.	0.6051 × 85.27	3 places	______	______
4.	74.326 × 0.34	4 places	______	______
5.	2.8703 × 9.425	4 places	______	______

Check your work on page 63. Record your score on page 67.

AMOUNTS OF MONEY IN MULTIPLICATION

Calculators make multiplying amounts of money easy because the calculator places the decimals for you. Just press the [·] Decimal Point Key in all factors where there is a decimal. Enter **6.7** × **3.21** = into your calculator. The product, 21.507, should be in the answer window.

Many business and personal math problems are about money (dollars and cents) and have only two decimal places. When you multiply amounts of money, you need only two decimal places in the product. If the product in the answer window has more than two decimal places, round to get a number that has two decimal places.

To round $9.43 × 7.6 = $71.668:

1. Count three places to the right of the decimal (8).
2. If the digit is
   a. *5 or more,* add 1 to the number to the left of that digit (8 is *5 or more,* so add 1 + 6 = 7).
   b. *less than 5,* keep the same digit in the place to the left.
3. The rounded product is $71.67.

CHECKPOINT 4-4

YOUR GOAL: Get 4 of 5 answers correct.

Find the product. Round it to two decimal places.

- Jill fills her car with 12.5 gallons of gasoline. The price of gasoline is $1.309. How much will Jill pay for the gasoline?

  $16.17

1. At the supermarket you buy 4.2 pounds of meat. The price of meat is $3.17 a pound. How much will you pay?

   ______

2. Troy's bedroom needs painting. Paint costs $15.95 a gallon. He needs four gallons of paint. How much will Troy pay for the paint?

   ______

3. Jack makes $4.25 an hour. He worked 36.5 hours. How much will Jack be paid?

   ______

4. Roger wants to buy new tires for his car. One new tire costs $89.95. Roger buys four tires. How much will he pay?

   ______

5. Dana Brown buys 12 calves from a livestock sale. Each calf costs $450. How much will farmer Brown pay?

   ______

Check your work on page 63. Record your score on page 67.

ESTIMATING PRODUCTS

Estimating multiplication products helps you know if the product seems correct.

For example, you estimate that a product should be near 600, but you get 6,390 for an answer. You know that you either estimated wrong or calculated wrong. Rework the problem to get the correct answer.

To estimate the product for 64 × 9:

1. Round the factors to the highest place at the left (64 becomes 60, and 9 becomes 10). Write the rounded factors in the Rounded Factors column.
2. Multiply the rounded factors on your calculator. Write the product (600) in the Estimated Product column.
3. Use your calculator to multiply the original factors (64 × 9). Write the total (576) in the Calculator Product column.
4. If the Calculator Product is not close to the Estimated Product, estimate and calculate the answer again. Write the correct Estimated Product and the Calculator Product.

CHECKPOINT 4-5

YOUR GOAL: Get 12 of 15 answers correct.

Round the factors. Find the Estimated Products and the Calculator Products.

	Original Factors	Rounded Factors	Estimated Product	Calculator Product
●	64 × 9	60 × 10	600	576
1.	86 × 32	______	______	______
2.	395 × 61	______	______	______
3.	457 × 308	______	______	______
4.	4,067 × 48	______	______	______
5.	8,962 × 409	______	______	______

Check your work on page 63. Record your score on page 67.

JUST FOR FUN

Write the number answer in the Answer column. Turn your calculator upside down and write the word in the Word column.

		Answer	Word
1.	3,869 × 2 =	______	______
2.	169 × 2 =	______	______
3.	551 × 14 =	______	______
4.	151 × 4 =	______	______

WHAT YOU HAVE LEARNED

Now that you have completed this unit, you should be able to:

- Multiply whole numbers correctly.
- Use clues to find multiplication word problems.
- Estimate multiplication answers.
- Multiply amounts of money correctly.
- Round decimal answers in multiplication problems.

PUTTING IT TOGETHER

ACTIVITY 4-1 YOUR GOAL: Get 4 of 5 answers correct.

Find the products.

- ● 73 × 51 = 3,723
1. 82 × 65 = ______
2. 821 × 57 = ______
3. 1,840 × 5,297 = ______
4. 976 × 51 = ______
5. 6,439 × 704 = ______

Check your work on page 63. Record your score on page 67.

ACTIVITY 4-2 YOUR GOAL: Get 4 of 5 answers correct.

Find the products.

- ● John drives 18 miles to work and back home every day. He traveled to work 23 days in April. How many miles did John travel to work and back home in April?

  414

1. Henry feeds his cattle seven bags of grain a week. Each bag weighs 50 pounds. How many pounds of grain does Henry feed a week?

   ______

2. Paula employs 12 people. Each person works 40 hours a week. How many hours do Paula's employees work each week?

   ______

3. Randall Recycling volunteers collect newspaper in bundles. They put 15 pounds of paper in each bundle. The volunteers collected 124 bundles during one drive. How many pounds of paper did the volunteers collect?

   ______

4. Mike used 25 boxes of screws while working on his construction job. Each box has 150 screws. How many screws did Mike use?

   ______

5. The community theater sold out all six nights of its performances. The theater seats 525 people. How many people saw the play?

______

Check your work on page 63. Record your score on page 67.

ACTIVITY 4-3 YOUR GOAL: Get 8 of 10 answers correct.

Find the Unrounded Product. Round it to the number of places in the Round To column.

		Round To	Unrounded Product	Rounded Product
●	48.9 × 0.33	2 places	16.137	16.14
1.	2.04 × 0.052	2 places	______	______
2.	0.136 × 0.457	3 places	______	______
3.	7.1916 × 44.5	3 places	______	______
4.	30.973 × 4.02	4 places	______	______
5.	70.3084 × 4.2	4 places	______	______

Check your work on page 63. Record your score on page 67.

ACTIVITY 4-4 YOUR GOAL: Get 4 of 5 answers correct.

Find the product. Round it to two decimal places.

● Gasoline is $1.329 per gallon. You put 13.5 gallons of gasoline in your car. How much will you owe?

$17.94

1. Keisha bought 16 loaves of bread for a fund-raising supper. Each loaf costs $1.36. How much did Keisha spend on bread?

______

2. Pat wants to put tile on her kitchen floor. The tiles cost $7.65 a square yard. Pat needs nine square yards. How much will Pat pay?

______

3. Carlos earns $5.25 per hour. He worked 37.25 hours. How much did he earn?

____

4. Darin needs to buy home heating oil. Oil costs $1.32 per gallon. He buys 252.5 gallons. How much does Darin owe?

____

5. Mohan Movie Theater seats 175 people. On Friday night all of the seats were filled. Each seat costs $5. How much did the movie theater collect?

____

☞ *Check your work on page 63. Record your score on page 67.*

ACTIVITY 4-5 YOUR GOAL: Get 12 of 15 answers correct.

Give the Rounded Factors, the Estimated Products, and the Calculator Products.

		Rounded Factors	Estimated Product	Calculator Product
●	88 × 23	90 × 20	1,800	2,024
1.	74 × 96			
2.	510 × 87			
3.	375 × 224			
4.	1,852 × 119			
5.	7,485 × 4,261			

☞ *Check your work on page 63. Record your score on page 67.*

MAKING IT WORK

REVIEW 4-1 YOUR GOAL: Get 5 of 6 answers correct.

Inventory

When a person wants to know how many items he or she has to sell or use, an **inventory** of items on hand is taken. An inventory is a complete list of the number and type of each item in stock. For example:

Fred sells fresh fruit and vegetables. He decided to take inventory of his produce. He filled in the following chart.

Find out how many of each fruit and vegetable Fred has. Multiply the number of baskets of fruit (5 baskets of apples) by the number of pieces of fruit in each basket (12 apples per basket).

Fred's Fresh Produce Date: August 7, 1992

Vegetable/Fruit	Number per Basket	Number of Baskets	Total
Apples	12	5	60
Tomatoes	12	15	______
Cucumbers	4	11	______
Carrots	12	9	______
Cantelope	2	19	______
Corn	6	21	______
Onions	12	17	______

Check your work on page 63. Record your score on page 67.

UNIT 5

Division—Using It Every Day

WHAT YOU WILL LEARN

When you finish this unit, you will be able to:

- Get correct answers to division problems.
- Use clues to find division word problems.
- Round to get answers to division problems with decimals.
- Estimate answers to division problems.
- Get correct answers to division money problems.
- Get correct answers to chain division problems.

THE BASICS OF DIVISION

This is a division problem:

Quotient → 5

$3\overline{)15}$ ← Dividend

↑

Divisor

The **dividend** (15) is the number to be divided. The **divisor** (3) is the number that does the dividing. The **quotient** (5) is the answer to a division problem.

Writing Division Problems

There are three ways to write a division problem:

$\text{Divisor}\overline{)\text{Dividend}}$ with Quotient on top $3\overline{)15}$ with 5 on top	Dividend ÷ Divisor = Quotient 15 ÷ 3 = 5	$\frac{\text{Dividend}}{\text{Divisor}}$ = Quotient $\frac{15}{3} = 5$

In each case, always enter the dividend first. Then enter the divisor into your calculator.

Calculator Division

To solve the problem 92 ÷ 4 = 23:

1. Enter the dividend, 92.
2. Press the ÷ Division Key. In your mind, say "divided by."
3. Enter the divisor, 4.
4. Press the = Equal Key.
5. The quotient, 23, should be in the answer window.

   If you did not get 23, work the problem until you get 23. Read each step slowly.
6. Clear the calculator.

Prove the Quotient

To prove a division quotient, multiply the quotient by the divisor. For example:

23 × 4 = 92

If the answer is the same as the dividend, the quotient is correct. If the answer is different, work the problem again until the proof answer is the same as the dividend.

CHECKPOINT 5-1

YOUR GOAL: Get 4 of 5 answers correct.

Find the quotients. Prove your answers. Write them in the blank spaces.

- 108 ÷ 12 = 9 Proof: 9 × 12 = 108

1. 338 ÷ 13 = ______
2. 2,730 ÷ 15 = ______
3. 10,404 ÷ 68 = ______
4. 8,432 ÷ 34 = ______
5. 12,194 ÷ 469 = ______

Check your work on page 63. Record your score on page 67.

SOLVING DIVISION WORD PROBLEMS

Division is finding out how many equal parts are in a whole. Division takes a whole item, such as a piece of cloth, and "cuts" it into equal parts.

The clue to a division problem is found by looking for a whole that needs to be cut into equal parts.

You picked 36 apples from a tree in your yard. You decide to bring them to class and divide them between your classmates. There are 12 students in your class. You want to give each classmate the same number of apples. How many apples will each classmate get?

This is a division problem because a number (the dividend) is to be divided by another number (the divisor).

The trick to solving a division word problem is figuring out which is the dividend and which is the divisor. After you find the dividend and the divisor, you can complete the division problem and find the quotient. For example, in the apple problem the dividend is the item you are going to divide up: the apples! The divisor is the number of people who will get the apples. The quotient is the number of apples each classmate will get. From THE BASICS OF DIVISION, you know that:

Dividend ÷ Divisor = Quotient

so:

36 apples ÷ 12 classmates = *3 apples per person*

CHECKPOINT 5-2

YOUR GOAL: Get 4 of 5 answers correct.

Read the following problems. Find the clues that tell you if it is a division problem.

If it is a division problem, write the answer in the blank space. If it is not a division problem, write "No ÷ " in the answer space.

- Will's Window Cleaning Service has to clean the windows on the local bank. There are 960 windows on the bank building. Will has 15 employees. How many windows will each employee have to clean?

64

1. Jimmy has to plow a 30-acre field. He decides he can plow five acres an hour. How many hours will it take for Jimmy to plow the field?

____

2. Penny buys six cans of juice for $1.06 each. How much will Penny pay for the juice?

____

3. Kelvin drove 274 miles. He then filled up the tank with 13.7 gallons of gasoline. How many miles a gallon does Kelvin's car get?

______

4. There are three boxes of fruit in the storeroom. One box has five cans. One box has seven cans. The other box has 13 cans. How many cans of fruit are in the storeroom?

______

5. A road crew needs to fix 2,000 feet of road. They want to put the caution barrels five feet apart. How many caution barrels does the road crew need?

______

Check your work on page 63. Record your score on page 67.

CALCULATORS MAKE IT EASY!

Calculators make division easier than doing the problem by pencil and paper. On a calculator you do not have to do as many steps to get an answer. For example:

To work the problem 1,965 ÷ 5, you have to do these steps to get the answer.

By Pencil and Paper

```
   393
5)1965   Write the numbers
  15        on paper.
   46    You do the
   45       work in your
    15      head and by
    15      hand.
     0
```

By Calculator

```
   393
5)1965   Press the numbers,
            and the calculator
            does the work
            automatically.
```

ROUNDING UNROUNDED QUOTIENTS

Quotients with decimals can be unrounded. This means the answer will have as many decimal places as appear in the answer window.

For example:

1. Enter the problem.

   65 ÷ 8 = 8.125

2. The quotient, 8.125, is an unrounded quotient (8.125 has three decimal places).

An unrounded quotient may be more exact than you need.

To round 8.125 to the nearest whole number, follow the instructions in Unit 2.

The rounded quotient is 8.13.

CHECKPOINT 5-3

YOUR GOAL: Get 8 of 10 answers correct.

Find the Unrounded Quotient. Round each number to the number of places in the Round To column.

	Round To	Unrounded Quotient	Rounded Quotient
● 6.58 ÷ 4	2 places	1.645	1.65
1. 79 ÷ 8	2 places	______	______
2. 96.3 ÷ 8	3 places	______	______
3. 339 ÷ 7	3 places	______	______
4. 257 ÷ 14	4 places	______	______
5. 452 ÷ 102	4 places	______	______

☞ *Check your work on page 64. Record your score on page 67.*

ESTIMATING QUOTIENTS

Estimating division quotients helps you know if the answer seems correct.

For example, if you estimate that a quotient should be near 600 but you get 62 for an answer, you know you either estimated wrong or calculated wrong. Rework the problem to get the correct answer.

To estimate the quotient for 294 ÷ 21:

1. Round the numbers to the highest place at the left (294 becomes 300, and 21 becomes 20). Write the rounded factors in the Rounded Factors column.
2. Divide the rounded numbers on your calculator. Write the quotient (15) in the Estimated Quotient column. Round to the nearest whole number.
3. Use your calculator to divide the original numbers (294 ÷ 21). Write the quotient (14) in the Calculator Product column.
4. If the Calculator Product is not close to the Estimated Product, estimate and calculate the answer again. Write the correct Estimated Product and the Calculator Quotient.

CHECKPOINT 5-4

YOUR GOAL: Get 12 of 15 answers correct.

Round the numbers. Find the Estimated Quotients and the Calculator Quotients.

	Original Numbers	Rounded Numbers	Estimated Quotient	Calculator Quotient
●	294 ÷ 21	300 ÷ 20	15	14
1.	192 ÷ 24			
2.	1,768 ÷ 52			
3.	2,856 ÷ 34			
4.	4,564 ÷ 326			
5.	32,660 ÷ 1,420			

Check your work on page 64. Record your score on page 67.

AMOUNTS OF MONEY IN DIVISION

Calculators automatically place a decimal at the correct place in a quotient. This makes dividing amounts of money easy.

Many business and personal math problems are about money (dollars and cents) that has only two decimal places. When you divide amounts of money, you need only two decimal places in the quotient. If the quotient in the answer window has more than two decimal places, round to two decimal places.

To round $1.98 ÷ 32 = 0.061875, follow the instructions in Unit 2.

The rounded quotient is $0.06.

CHECKPOINT 5-5

YOUR GOAL: Get 4 of 5 answers correct.

Find the quotient. Round it to two decimal places.

● Apples cost $.70 a pound. How many pounds can you buy with $3.50? **5**

1. Shampoo costs $.98. There are 22 ounces in one bottle of shampoo. How much does the shampoo cost an ounce? ______

2. Martin brought 2,500 bushels of wheat to the market. He received $5,250 for the wheat. How much money did Martin get per bushel for his wheat?

____

3. Four friends rent an apartment. The rent is $525 a month. How much will each friend have to pay per month?

____

4. Gasoline costs $1.349 a gallon. Karmen's car gets 22 miles to the gallon. How much does gasoline cost per mile for Karmen?

____

5. Charles bought eight spark plugs. He wrote the cost of $15.84 in his checkbook. Jim asked Charles to sell him a spark plug. How much should Charles charge Jim for one spark plug?

____

Check your work on page 64. Record your score on page 67.

CHAIN DIVISION

Your calculator makes it possible to divide a quotient by another divisor one or more times. This is called **chain division.**

To work: 126 ÷ 24 ÷ 6 = 0.88

1. Enter the dividend (126), and press [÷].
2. Enter the divisor (24), and press [÷].
3. Enter the second divisor (6), and press [=].
4. Round the answer (0.875) to two decimal places.
5. Write the rounded quotient (0.88) in the blank space.

CHECKPOINT 5-6

YOUR GOAL: Get 4 of 5 answers correct.

Use chain division to solve the problems. Round the answer to two decimal places.

●	728 ÷	7 ÷	8 =	13
1.	705 ÷	20 ÷	25 =	________
2.	1,344 ÷	48 ÷	15 =	________
3.	13,167 ÷	627 ÷	3 =	________

4. 386,274 ÷ 119 ÷ 14 = ________

5. 554,268 ÷ 34 ÷ 627 = ________

Check your work on page 64. Record your score on page 67.

JUST FOR FUN

Work the problem and turn the calculator upside down to read the answer to the question.

1. What does a person do when he or she has a broken leg?

   1,515,216 ÷ 4 = ________
2. Sometimes your blood does this.

   42,648 ÷ 6 = ________
3. If a Thanksgiving dinner could talk, what would it say?

   757,612 ÷ 2 = ________
4. This is what you wear over your sock.

   45,675 ÷ 15 = ________
5. You do not want to fall into one of these.

   29,632 ÷ 8 = ________

WHAT YOU HAVE LEARNED

Now that you have completed this unit, you should be able to:

- Divide whole numbers correctly.
- Use clues to find division word problems.
- Round decimal answers in division problems.
- Estimate division answers.
- Divide amounts of money correctly.
- Solve chain division problems.

PUTTING IT TOGETHER

ACTIVITY 5-1 YOUR GOAL: Get 4 of 5 answers correct.

Find the quotients. Do not round.

- ● 42 ÷ 8 = 5.25
1. 85 ÷ 4 = ______
2. 307 ÷ 8 = ______
3. 568 ÷ 16 = ______
4. 511 ÷ 73 = ______
5. 26,214 ÷ 771 = ______

Check your work on page 64. Record your score on page 67.

ACTIVITY 5-2 YOUR GOAL: Get 4 of 5 answers correct.

Find the quotients.

● Emitt feeds his cattle 35,000 pounds of grain a week. He feeds the same amount of grain each day. How many pounds of grain does Emitt feed his cattle each day?

5,000

1. Christine is making chicken nuggets at Casy's Restaurant. A serving is six nuggets. She cooks a case of nuggets. One case has 480 nuggets in it. How many servings are prepared for sale?

______

2. Marcy puts cheese in her sandwiches for work each day. She buys a package of cheese that has 75 slices in it. Marcy uses 1.5 pieces of cheese a day. How many days will Marcy have cheese for her sandwiches?

______

3. The Cleaning Service has 525 rags for their workers to use. There are 25 workers on the Cleaning Service staff. How many rags will each worker get to use?

______

4. Randy made six trips to work and back home. He has gone 102 miles. How far is Randy's trip to work and back home?

____

5. Toby needs 675 screws to finish a shop job. Each box has 75 screws in it. How many boxes does Toby need to finish the job?

____

Check your work on page 64. Record your score on page 67.

ACTIVITY 5-3 YOUR GOAL: Get 12 of 15 answers correct.

Give the Estimated Quotients and the Calculator Products.

	Original Numbers	Rounded Numbers	Estimated Quotient	Calculator Quotient
●	95 ÷ 5	100 ÷ 10	10	19
1.	235 ÷ 5	________	________	________
2.	611 ÷ 47	________	________	________
3.	5,706 ÷ 9	________	________	________
4.	3,345 ÷ 15	________	________	________
5.	49,956 ÷ 724	________	________	________

Check your work on page 64. Record your score on page 67.

ACTIVITY 5-4 YOUR GOAL: Get 4 of 5 answers correct.

Find the quotient. Round it to two decimal places.

● George works at a carpentry shop. He and two other workers built a cabinet. They sold the cabinet and had a profit of $96. How much money will each of the three get for the cabinet?

$32

1. Kami bought seven square yards of carpet for a room in her home. Her husband wants to know how much she paid per square yard for the carpet. She knows she paid $68.95 for all of the carpet. How much did she pay per square yard?

____

2. Crystallia worked 37 hours last week. She got paid $175.75. How much does Crystallia earn per hour?

______

3. Amanda wants to buy her four children small rewards for getting good grades in school. She only has $10 to spend. How much can she spend on each child?

______

4. Kirk delivered 36,500 pounds of tomatoes to the produce terminal. If the tomatoes were packed in buckets of 50 pounds each, how many buckets of tomatoes were delivered?

______

5. Melissa puts a microwave oven on layaway. After she pays the down payment, she owes $276.82. She plans to pay it off in eight months. How much will she have to pay each month?

______

☞ *Check your work on page 64. Record your score on page 67.*

ACTIVITY 5-5 YOUR GOAL: Get 8 of 10 answers correct.

Find the Unrounded Quotient. Round each number to the number of places in the Round To column.

		Round To	Unrounded Quotient	Rounded Quotient
●	34 ÷ 16	2 places	2.125	2.13
1.	108.9 ÷ 12	2 places	______	______
2.	572 ÷ 6	3 places	______	______
3.	6.75 ÷ 4	3 places	______	______
4.	484 ÷ 73	4 places	______	______
5.	896 ÷ 405	4 places	______	______

☞ *Check your work on page 64. Record your score on page 67.*

ACTIVITY 5-6 YOUR GOAL: Get 4 of 5 answers correct.

Use chain division to solve the problems.

● 291 ÷ 30 ÷ 2 = 4.85

1. 810 ÷ 18 ÷ 5 = ______

2. 276 ÷ 24 ÷ 23 = ______

3. 6,844 ÷ 58 ÷ 2 = ______

4. 5,375 ÷ 43 ÷ 25 = ______

5. 27,648 ÷ 36 ÷ 3 = ______

☞ *Check your work on page 64. Record your score on page 67.*

CHECKING WHAT YOU LEARNED

Take this posttest after completing *Basic Math With A Calculator.* The 30 questions will help you tell what skills you have mastered. They will also give you an idea of those skills you should review. There is no time limit, so take the time you need.

Follow the directions for each section. Write your answers in the space provided. When you finish, check your work against the answers near the back of the book. Give yourself 1 point for each correct answer. Record your score in your Personal Progress Record. The analysis chart will help tell where you may need additional study.

DIRECTIONS: Use your calculator to find the answers to all of the problems below. Write your answers in the blank spaces.

(1) Find the total. 389 + 76 + 7,425 = **1.** ______

Column A shows the correct number. Draw a line through the number in Column B if it is different from the number in Column A. Write in the blank space the correct number from Column A.

Column A	Column B
(2) $4,817.29	$4.817.20

2. ______

(3) Round 257 to the nearest hundreds place. **3.** ______

Work the following problem: 528 + 435

(4) Rounded Problem **(5)** Estimated Total **(6)** Calculator Total

4. ______

5. ______

6. ______

Underline the clues to these addition problems. Then work the problems to find the correct totals.

(7) Brenda is in a weight-reducing class. In three months, she lost 24 pounds, 17 pounds, and 10 pounds. How many pounds did she lose all together? **7.** ______

(8) Allen wants to estimate the miles to his aunt's house. The map says the distance is 476 miles. Round these miles to hundreds. **8.** ______

(9) Find the total. **9.** ______

```
  $24.27
     .62
   28.53
   79.06
 + 30.94
```

(10) Find the difference. 235 − 260 = **10.** ______

Estimate and calculate the differences for this problem: 268 − 47 =

(11) Rounded Problem **11.** ______

(12) Estimated Difference **12.** ______

(13) Calculator Difference **13.** ______

Underline the clue that tells you to subtract in this word problem. Find the difference.

(14) Marcia's grocery bill total came to $8.41. She gave the cashier $10. How much money does she have left over? **14.** ______

(15) Find the difference. $459.37 − 39.84 = **15.** ______

(16) Find the product. 863 × 42 = **16.** ______

(17) The community theater sold out all 9 nights of its performances. The theater seats 250 people. How many people saw the play? **17.** ______

(18) Round 8.4057 to hundredths. **18.** ______

Estimate and calculate the products for this problem: 645 × 78 =

(19) Rounded Problem **19.** ______

(20) Estimated Product **20.** ______

(21) Calculator Product **21.** ______

(22) Find the quotients. Round to hundredths. 423 ÷ 62 = **22.** ______

(23) Lowell made 6 trips to work and back home. He traveled 96 miles. How far is Lowell's trip to work and back home? **23.** ______

Find the Unrounded Quotient. Then find the Rounded Quotient. Round to two decimal places. 468 ÷ 35 =

(24) Unrounded Quotient **24.** ______

(25) Rounded Quotient **25.** ______

Estimate and calculate the quotients for this problem. Round the calculator answer to two decimal places. 526 ÷ 38 =

(26) Rounded Problem **26.** ______

(27) Estimated Quotient **27.** ______

(28) Calculator Quotient **28.** ______

Find the quotient. Round to two decimal places.

(29) Laurie worked 28 hours last week. She got paid $198.57. How much does Laurie earn an hour? **29.** ______

Use chain division to solve this problem. Round your answer to two decimal places.

(30) 560 ÷ 35 ÷ 72 **30.** ______

☞ ***Check your answers on page 64. Record your score on page 68.***

GLOSSARY

A

Addends Two or more numbers to be added.
Addition Combining two or more numbers to get a total.
Align To place in a straight line.

C

Chain division Dividing a quotient by another divisor in a continuous process.

D

Difference The answer to a subtraction problem.
Digit One of the numbers 0, 1, 2, 3, 4, 5, 6, 7, 8, or 9.
Dividend The number to be divided in a division problem.
Divisor The number that the dividend is divided by.

E

Estimate An approximate calculation.

F

Factors The multiplicand and the multiplier in a multiplication problem.

H

Horizontal Across.

I

Inventory A complete list of the number and type of each item in stock.

L

Legibly Clearly.

M

Minuend The number being subtracted from.
Multiplicand The number in a multiplication problem which is added to itself a given number of times.
Multiplication Adding the multiplicand to itself the number of times given in the multiplier.
Multiplier The number in a multiplication problem which tells how many times to add the multiplicand to itself.

P

Product The answer to a multiplication problem.
Proofread (numbers) To compare each digit in a number with digits in another number to determine if they are identical.
Prove the answer Work the problem a second time in a different way.

Q

Quotient The answer to a division problem.

R

Round To change a number to the nearest power of ten.

S

Subtraction Deducting one number from another number.
Subtrahend The number being subtracted.
Sum The answer to an addition problem; also called *total*.

T

Total The answer to an addition problem; also called *sum*.

V

Vertical Up and down.

W

Word problem A math problem described in a story or scenario.

INDEX

ANSWERS

✔ CHECKING WHAT YOU KNOW

1. 6,773
2. $3,928.18
3. 400
4. 400 + 500
5. 900
6. 965
7. all together 29
8. 200
9. $77.33
10. 67
11. 200 − 60
12. 140
13. 131
14. left over $.42
15. $309.33
16. 24,882
17. 4,445
18. 7.39
19. 500 × 90
20. 45,000
21. 46,814
22. 7.32
23. 21
24. 23.833333
25. 23.83
26. 600 ÷ 50
27. 12
28. 12.96
29. $7.33
30. .43

UNIT 1

CHECKPOINT 1-1, page 3

1. 83
2. 694
3. 372
4. 539
5. 970

CHECKPOINT 1-2, page 4

1. seven two
2. three seven one
3. six nine eight
4. eight [pause] nine six four
5. two [pause] five one four

CHECKPOINT 1-3, page 5

Answers will vary.

CHECKPOINT 1-4, page 7

Column C

1. 6431 ÷ 839
2. 11:08 a.m.
3. 214-401-5970
4. 8/14/88
5. 368907

ACTIVITY 1-1, page 9

1. 1,628
2. 1,637
3. 8,959
4. 99,772
5. 29,383

ACTIVITY 1-2, page 9

1. $1,892.43
2. 12:38 p.m.
3. SQ-178-3062
4. 9/24/93
5. 15973628

UNIT 2

CHECKPOINT 2-1, page 11

1. 330
2. 5,100
3. 3,000
4. 7,600
5. 5,000

CHECKPOINT 2-2, page 12

	Rounded Addends	Est. Total	Calc. Total
1.	20 + 40	60	58
2.	30 + 700	730	774
3.	300 + 400	700	697
4.	2,000 + 5,000	7,000	7,222
5.	2,000 + 50,000 + 7,000	59,000	55,781

CHECKPOINT 2-3, page 14

1. all together 33
2. No +
3. total $7,230
4. No +
5. how much $30

CHECKPOINT 2-4, page 16

1. 43.52
2. $101.61
3. $740.59
4. 151.923
5. 45.7249

ACTIVITY 2-1, page 18

1. 30
2. 80
3. 800
4. 4,000
5. 9,000

ACTIVITY 2-2, page 18

	Rounded Addends	Est. Total	Calc. Total
1.	40 + 50	90	89
2.	50 + 200	250	244
3.	600 + 200	800	855
4.	800 + 900	1,700	1,723
5.	100 + 600	700	701

ACTIVITY 2-3, page 18

1. in all 1,030
2. 400
3. all together 100
4. total 139
5. total $3.40

ACTIVITY 2-4, page 19

1. 79.45
2. $59.68
3. $48.31
4. 1,323.2719
5. 727.186

UNIT 3

CHECKPOINT 3-1, page 22

1. 21 Proof: 21 + 14 = 35
2. 45 Proof: 45 + 34 = 79
3. 254 Proof: 254 + 89 = 343
4. 331 Proof: 331 + 231 = 562
5. 6,574 Proof: 6,574 + 856 = 7,430

CHECKPOINT 3-2, page 24

	Rounded Problem	Estimated Difference	Calculator Difference
1.	80 − 40	40	39
2.	900 − 80	820	786
3.	200 − 100	100	92
4.	400 − 200	200	205
5.	300 − 100	200	138

CHECKPOINT 3-3, page 25

1. difference 8
2. less 194
3. reduces $0.56
4. remains $9
5. left 7

CHECKPOINT 3-4, page 26

1. 56.30
2. 25.1054
3. 10.161
4. 1633.371
5. 144.18

CHECKPOINT 3-5, page 27

1. 5.37
2. $24.95
3. 9.109
4. $10.56
5. 52.777

JUST FOR FUN, page 27

1. 57,716 GILLS
2. 77,345,663 EGGSHELL
3. 38,076 GLOBE
4. 5,514 HISS
5. 818 BIB

ACTIVITY 3-1, page 29

1. 35 Proof: 35 + 54 = 89
2. 359 Proof: 359 + 96 = 455
3. 75 Proof: 75 + 73 = 148
4. 639 Proof: 639 + 36 = 675
5. 3331 Proof: 3331 + 245 = 3576

ACTIVITY 3-2, page 29

	Rounded Problem	Estimated Difference	Calculator Difference
1.	50 − 20	30	31
2.	200 − 50	150	188
3.	500 − 100	400	372
4.	200 − 60	140	122
5.	400 − 300	100	118

ACTIVITY 3-3, page 30

1. difference $56
2. change $0.27
3. less $75
4. left over $2.64
5. remains $169.31

ACTIVITY 3-4, page 30

1. 19.451
2. $338.55
3. 59385.673
4. $60.97
5. 340.375

REVIEW 3-1, page 31

Pants	$18.04
Socks	$2.81
Shirt	$4.74
Sweater	$5.50

Totals

Regular Price	$134.35
Sale Price	$95.86
Difference	$38.49

REVIEW 3-2, page 32

Bread	$1.19
Dog Food	$7.24
Milk	$2.40
Ground Beef	$1.88

Totals

Regular Price	$18.09
Less Coupon	$3.88
Cost with Coupon	$14.21

UNIT 4

CHECKPOINT 4-1, page 34

1. 294
2. 59,644
3. 1,371,905
4. 2,945
5. 227,752

CHECKPOINT 4-2, page 35

1. 135
2. No ×
3. 405
4. No ×
5. 124

CHECKPOINT 4-3, page 37

	Unrounded Product	Rounded Product
1.	59.5886	59.59
2.	5.912286	5.912
3.	51.596877	51.597
4.	25.27084	25.2708
5.	27.052577	27.0526

CHECKPOINT 4-4, page 38

1. $13.31
2. $63.80
3. $155.13
4. $359.80
5. $5,400.00

CHECKPOINT 4-5, page 39

	Rounded Factors	Estimated Product	Calculator Product
1.	90 × 30	2,700	2,752
2.	400 × 60	24,000	24,095
3.	500 × 300	150,000	140,756
4.	4,000 × 50	200,000	195,216
5.	9,000 × 400	3,600,000	3,665,458

JUST FOR FUN, page 39

1. 7,738	Bell
2. 338	Bee
3. 7,714	Hill
4. 604	Hog

ACTIVITY 4-1, page 41

1. 5,330
2. 46,797
3. 9,746,480
4. 49,776
5. 4,533,056

ACTIVITY 4-2, page 41

1. 350
2. 480
3. 1,860
4. 3,750
5. 3,150

ACTIVITY 4-3, page 42

	Unrounded Product	Rounded Product
1.	0.10608	0.11
2.	0.062152	0.062
3.	320.0262	320.026
4.	124.51146	124.5115
5.	295.29528	295.2953

ACTIVITY 4-4, page 42

1. $21.76
2. $68.85
3. $195.56
4. $333.30
5. $875.00

ACTIVITY 4-5, page 43

	Rounded Factors	Estimated Product	Calculator Product
1.	70 × 100	7,000	7,104
2.	500 × 90	45,000	44,370
3.	400 × 200	80,000	84,000
4.	2,000 × 100	200,000	220,388
5.	7,000 × 4,000	28,000,000	31,893,585

REVIEW 4-1, page 44

Tomatoes, 180	Cantelope, 38
Cucumbers, 44	Corn, 126
Carrots, 108	Onions, 204

UNIT 5

CHECKPOINT 5-1, page 46

1. 26
2. 182
3. 153
4. 248
5. 26

CHECKPOINT 5-2, page 47

1. 6
2. No ÷
3. 20
4. No ÷
5. 400

CHECKPOINT 5-3, page 49

	Unrounded Quotient	Rounded Quotient
1.	9.875	9.88
2.	12.0375	12.038
3.	48.428571	48.429
4.	18.357142	18.3571
5.	4.4313725	4.4314

CHECKPOINT 5-4, page 50

	Rounded Numbers	Est. Quo.	Calc. Quo.
1.	200 ÷ 20	10	8
2.	2,000 ÷ 50	40	34
3.	3,000 ÷ 30	100	84
4.	5,000 ÷ 300	17	14
5.	30,000 ÷ 1,000	30	23

CHECKPOINT 5-5, page 50

1. $0.04
2. $2.10
3. $131.25
4. $0.06
5. $1.98

CHECKPOINT 5-6, page 51

1. 1.41
2. 1.87
3. 7
4. 231.86
5. 26

JUST FOR FUN, page 52

1.	278,804	HOBBLE
2.	7,108	BOIL
3.	378,806	GOBBLE
4.	3,045	SHOE
5.	3,704	HOLE

ACTIVITY 5-1, page 53

1. 21.25
2. 38.375
3. 35.5
4. 7
5. 34

ACTIVITY 5-2, page 53

1. 80
2. 50
3. 21
4. 17
5. 9

ACTIVITY 5-3, page 54

	Rounded Numbers	Estimated Quotient	Calculator Quotient
1.	200 ÷ 10	20	47
2.	600 ÷ 50	12	13
3.	6,000 ÷ 10	600	634
4.	3,000 ÷ 20	150	223
5.	50,000 ÷ 700	71.43 or 71	69

ACTIVITY 5-4, page 54

1. $9.85
2. $4.75
3. $2.50
4. 730
5. $34.60

ACTIVITY 5-5, page 55

	Unrounded Quotient	Rounded Quotient
1.	9.075	9.08
2.	95.333333	95.333
3.	1.6875	1.688
4.	6.6301369	6.6301
5.	2.2123456	2.2123

ACTIVITY 5-6, page 56

1. 9
2. 0.5
3. 59
4. 5
5. 256

✔ CHECKING WHAT YOU LEARNED

1. 7,890
2. $4,817.29
3. 300
4. 500 + 400
5. 900
6. 963
7. all together 51
8. 500
9. $163.42
10. 197
11. 300 − 50
12. 250
13. 221
14. left over $1.59
15. $419.53
16. 36,246
17. 2,250
18. 8.41
19. 600 × 80
20. 48,000
21. 50,310
22. 6.82
23. 16
24. 13.371428
25. 13.37
26. 500 ÷ 40
27. 13
28. 13.84
29. $7.09
30. .22

PERSONAL PROGRESS RECORD

Name: ______________________________

✔ CHECKING WHAT YOU KNOW—ANALYSIS CHART

Use the chart below to determine the areas in which you may need additional study. In the space provided, write the total number of points you got for each content area. Each correct answer is worth 1 point. Add the total number of points to find your final score. Circle the items you answered incorrectly. Use the page references for those items to locate the topics. As you begin your study, pay close attention to those areas where you missed half or more of the questions.

Content Area	Item Number	Study Pages	Total Points	Number Right
Addition	1, 6, 7, 9	1, 2, 13–15	4	
Subtraction	10, 13, 14, 15	21, 22, 24–26	4	
Multiplication	16, 17, 21	33–35	3	
Division	22, 23, 24, 28, 29, 30	45–51	6	
Proofreading Numbers	2	7	1	
Rounding	3, 4, 8, 11, 18, 19, 25, 26	10, 11, 36, 48, 49	8	
Estimating	5, 12, 20, 27	12, 23, 38, 49	4	

Date ______________________ Total Points: 30 Your Score: ______

To record your score:

1. Count how many problems you have correct.
2. Record the number correct in the Score column.

UNIT 1 Addition—the Basics

Exercise	Score
Checkpoint 1-1	______
Checkpoint 1-2	______
Checkpoint 1-3	______
Checkpoint 1-4	______
Activity 1-1	______
Activity 1-2	______
Total, Unit 1	______

HOW ARE YOU DOING?

26 or better	Excellent
21–25	Good
19–20	Fair
Less than 19	See instructor

UNIT 2 Addition—Using It Every Day

Exercise	Score
Checkpoint 2-1	______
Checkpoint 2-2	______
Checkpoint 2-3	______
Checkpoint 2-4	______
Activity 2-1	______
Activity 2-2	______
Activity 2-3	______
Activity 2-4	______
Total, Unit 2	______

HOW ARE YOU DOING?

48 or better	Excellent
42–47	Good
36–41	Fair
Less than 36	See instructor

UNIT 3 Subtraction—Using It Every Day

Exercise	Score
Checkpoint 3-1	______
Checkpoint 3-2	______
Checkpoint 3-3	______
Checkpoint 3-4	______
Checkpoint 3-5	______
Activity 3-1	______
Activity 3-2	______
Activity 3-3	______
Activity 3-4	______
Review 3-1	______
Review 3-2	______
Total, Unit 3	______

HOW ARE YOU DOING?

56 or better	Excellent
46–55	Good
40–45	Fair
Less than 40	See instructor

UNIT 4 Multiplication—Using It Every Day

Exercise	Score
Checkpoint 4-1	______
Checkpoint 4-2	______
Checkpoint 4-3	______
Checkpoint 4-4	______
Checkpoint 4-5	______
Activity 4-1	______
Activity 4-2	______
Activity 4-3	______
Activity 4-4	______
Activity 4-5	______
Review 4-1	______
Total, Unit 4	______

HOW ARE YOU DOING?

56 or better	Excellent
49–55	Good
41–48	Fair
Less than 41	See instructor

UNIT 5 Division—Using It Every Day

Exercise	Score
Checkpoint 5-1	______
Checkpoint 5-2	______
Checkpoint 5-3	______
Checkpoint 5-4	______
Checkpoint 5-5	______
Checkpoint 5-6	______
Activity 5-1	______
Activity 5-2	______
Activity 5-3	______
Activity 5-4	______
Activity 5-5	______
Activity 5-6	______
Total, Unit 5	______

HOW ARE YOU DOING?

72 or better	Excellent
62–71	Good
54–61	Fair
Less than 54	See instructor

✔ CHECKING WHAT YOU LEARNED—ANALYSIS CHART

Use the chart below to determine the areas in which you may need additional study. In the space provided, write the total number of points you got for each content area. Each correct answer is worth 1 point. Add the total number of points to find your final score. Compare your score with your pretest score. Then circle the items you answered incorrectly. Use the page references for those items to locate the topics. Go back to those pages and review any items you may have missed.

Content Area	Item Number	Study Pages	Total Points	Number Right
Addition	1, 6, 7, 9	1, 2, 13–15	4	
Subtraction	10, 13, 14, 15	21, 22, 24–26	4	
Multiplication	16, 17, 21	33–35	3	
Division	22, 23, 24, 28, 29, 30	45–51	6	
Proofreading Numbers	2	7	1	
Rounding	3, 4, 8, 11, 18, 19, 25, 26	10, 11, 36, 48, 49	8	
Estimating	5, 12, 20, 27	12, 23, 38, 49	4	

Date ____________________

Total Points: 30 Your Score: ______